The POWER of Questioning

Guiding Student Investigations

Julie V. McGough and Lisa M. Nyberg

NSTApress

National Science Teachers Association

Arlington, Virginia

Dedicated to all teachers
who inspire children
with minds full of wonder
to seek answers to
a lifetime of questions.

National Science Teachers Association

Claire Reinburg, Director
Wendy Rubin, Managing Editor
Andrew Cooke, Senior Editor
Amanda O'Brien, Associate Editor
Donna Yudkin, Book Acquisitions Coordinator

ART AND DESIGN
Will Thomas Jr., Director

PRINTING AND PRODUCTION
Catherine Lorrain, Director

NATIONAL SCIENCE TEACHERS ASSOCIATION
David L. Evans, Executive Director
David Beacom, Publisher

1840 Wilson Blvd., Arlington, VA 22201
www.nsta.org/store
For customer service inquiries, please call 800-277-5300.

Library of Congress Cataloging-in-Publication Data
McGough, Julie V., 1969-
  The power of questioning : guiding student investigations / by Julie V. McGough and Lisa M. Nyberg.
    pages cm
Includes bibliographical references.
  ISBN 978-1-938946-28-8 (print) -- ISBN 978-1-941316-78-8 (e-book) 1. Science--Study and teaching. 2. Questioning. I. Nyberg, Lisa M., 1959- II. Title.
  LB1585.M375 2014
  507.1--dc23
                    2015001191

Cataloging-in-Publication Data for the e-book are available from the Library of Congress.

Contents

Contents

Color Coding

Throughout *The Power of Questioning*, the text, illustrations, and graphics are color-coded to indicate the components of the instructional model.

Questioning is printed in **red**.

Investigations are printed in **blue**.

Assessments are printed in **purple**.

When thoughtful **questioning** is combined with engaging **investigations**, amazing **assessments** are produced—just as when **red** and **blue** are combined, **purple** is produced.

We've also provided links and QR codes to the NSTA Extras page where you can view videos related to content throughout the book. Visit *www.nsta.org/publications/press/extras/questioning.aspx* to view all supplementary content.

Learn from yesterday, live for today, hope for tomorrow.

The important thing is not to stop questioning.

Albert Einstein
(Relativity: The Special and the General Theory, 1920)

PART 1
Why Is Questioning a Powerful Teaching Tool?

Connecting Questions and Learning

Why Is Questioning a Powerful Teaching Tool?

Children are full of questions every day: "Why? How does that work? Why? How long will that take? How do you know? Why?" The science classroom is the perfect place to take advantage of this natural curiosity. Questioning cultivates student engagement and drives instruction throughout the learning process.

The model we present here, Powerful Practices, takes questioning in the classroom to the next level while emphasizing the interconnected nature of instruction. You start with one question, but instead of having students look for the answer, you allow students to raise other questions as they investigate the topic further. Like ripples on a pond, questions follow one after another and provide new waves of instruction—from purposeful **questioning** to dynamic **investigations** to authentic **assessments**!

In *The Power of Questioning*, we invite you to explore the possibilities of questioning and experience the wonder of learning as you turn the pages of this pedagogical picture book.

Why Is Questioning Important When Linking Literacy to Learning Investigations and Authentic Performance Assessments?

A Framework for K–12 Science Education (*Framework*; NRC 2012) states, "Questions are the engine that drive science and engineering. … Asking questions is essential to developing scientific habits of mind" (p. 54). In the *Framework*, science and engineering are divided—the former to ask questions and the latter to define problems. Science questions may include, "How does it work? How can we find out how it works? How can we support what we know with evidence?" Engineering questions might include, "What is the problem? How can we solve the problem? What tools or supplies do we need to solve the problem? How can we support what we know with evidence?"

Questioning is a vital part of science and engineering. Questions challenge student thinking at key points within a lesson or unit of study to connect concepts and drive investigations. Student thinking deepens as the students integrate the skill of questioning into each lesson. As students interact with the world through questioning and collaboration with others, they build understanding. They learn to answer questions—and how to construct them. Together, through modeling and practice, the teacher and students develop the process of weaving questions throughout investigations and discussions.

Powerful Practices is an instructional model with questions at its core. The model guides students and teachers in the process of learning by emphasizing three key components: **questions**, **investigations**, to **assessments** (Figure 1.1). These three aspects are linked, and it is important not to approach the model in a linear fashion. You might start with a question or with an investigation.

Each component itself starts with a question because when learning experiences begin with engaging questions, they draw learners together through meaningful discussions and purposeful investigations. As students investigate a concept or phenomena, they ask questions, and questions help teachers assess student understanding.

Because it is multidimensional, the Powerful Practices model supports cross-curricular literacy development through science, technology, engineering, and mathematics (STEM).

Figure 1.1. The Powerful Practices Instructional Model
The three components—questions, investigations, and assessments—are interconnected throughout the unit of study.

How Does the Powerful Practices Instructional Model Work?

The Powerful Practices model is composed of three key components, each of which starts with a guiding question. The model emphasizes the interconnected nature of instruction and the importance of asking questions at every stage.

Questions

First, to engage learners in questioning, the teacher generates interest in the concept or phenomena by posing questions that stimulate thinking and discussion. For example, the teacher could begin a discussion about plants with, "What do you know about plants?" In the visual model of Powerful Practices, this starting question goes into the interior red circle (Figure 1.2). As the teacher and students share prior knowledge about plants, more questions develop, and possible misconceptions are revealed. Student or teacher questions that arise are added to outer red rings.

Investigations

The information gathered by the teacher through questioning leads to investigations. The teacher gathers resources (text, digital media, etc.) and plans observations and investigations for the students to perform at an *investigation station*. Continuing the example of studying plants, the investigation station could have real plants and simple plant experiments (e.g., planting seeds, putting celery stems in colored water, and examining leaves) for the students to explore. The initial question for this phase goes in the interior blue ring of the visual model (Figure 1.2). As the students investigate, their questions or wonderings go in the outer blue rings.

Assessments

The third, purple circle (Figure 1.2) in the visual model represents the assessments component. The inner purple ring may include a question to assess prior knowledge (e.g., "What do you know about plants?") or a question to frame the performance expectation found in the *Next Generation Science Standards* (*NGSS*; e.g., "What can we make or build that will help us communicate how a plant works?"). As students reflect on the learning process, they may generate more questions that drive new investigations to dig deeper (depth and complexity) or make connections to other concepts (crosscutting concepts) or other content areas (crosscurricular connections). These new questions form the outer rings of the purple assessment circle.

Figure 1.2. The Powerful Practices Model in Action
The inner circles show the initial question for each stage, and the outer rings show
questions that arise during discussions.

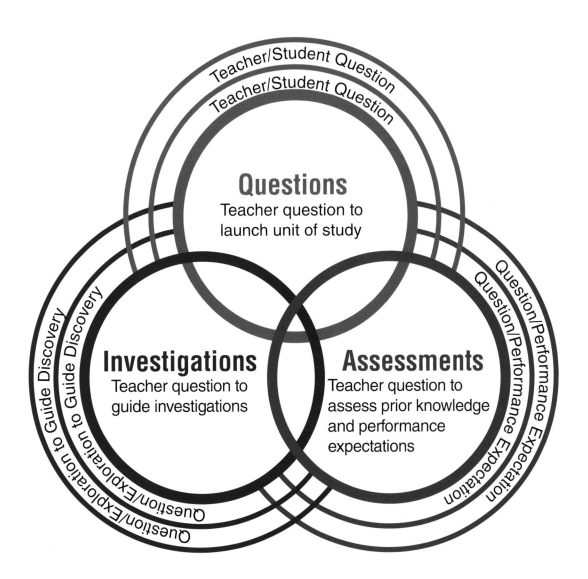

Why Does Skill in Questioning Engage Students in Purposeful Standards-Based Learning?

Students need opportunities to develop science literacy through solving problems and explaining phenomena and observations (NRC 2000). They also need to see purpose for what they are learning as they engage in literacy practices. Children ask questions and make connections to what is being learned in the classroom every day—on the playground, at home, walking to and from class, and when listening to stories and presentations. Sharing these connections through academic discourse helps students formulate new ideas and reconstruct old ones by adding new information from others' experiences.

Academic subjects are often regulated by national and state standards such as the *NGSS* and the *Common Core State Standards* (*CCSS*). These standards may lead teachers to engage children in higher-level thinking than they otherwise would through questioning, investigations, and authentic performance assessments. The standards build a bridge to connect real-world problem solving to the application of academic knowledge and skills. Additionally, the standards may guide teachers to engage children in complex cognitive processes so students may produce multidimensional work products illustrating higher-level thinking.

For example, during a study of the structure and function of plants, Cienna remembered her experience of noticing the tiny root hairs growing on a carrot while harvesting plants in the garden (photo on opposite page). She applied the information from the experience when building a model plant, deepening her understanding of the concept of how plant roots work. (Visit *www.nsta.org/publications/press/extras/files/practices/questioning/video2.htm* or scan the QR code on p. 18 to see a video.) Table 1.1 (p. 10) illustrates the *CCSS* and *NGSS* relevant to Cienna's discovery.

Table 1.1. Standards-Based Learning: Structure and Function of Plants

Examples of standards used during the study of the structure and function of plants. DOK = Depth of Knowledge (see p. 22); ELA, English language arts.

National Standards	Standards-Based Learning
NGSS: **Life Science** **LS1.A:** Structure and Function *NGSS*: **Engineering** **ETS1.2:** Developing and Using Models *CCSS ELA*: **Reading Informational Text** **RI.7:** Use illustrations and details in a text to describe and explain key ideas *CCSS ELA*: **Speaking and Listening** **SL.2:** Ask and answer questions about key details **SL.4:** Describe things with relevant details **CCR.4:** Present information, findings, and supporting evidence **SL.5:** Add visual displays to descriptions to clarify ideas	*NGSS* Children learn that plants have internal (xylem, phloem, veins) and external (roots, stems, leaves, flowers, fruits) parts that help them survive and grow by investigating (e.g., planting seeds, placing a carrot top in water) and observing real plants over time (e.g., garden experiences) (DOK Levels 1 and 2). Children develop models to describe phenomena (DOK Level 3). *CCSS ELA* Children ask questions about the parts of the plant and how the parts work to help the plant grow. The children use informational text to explain the different internal and external plant parts. Students describe how plants work and present their information to others using the model plant as a visual display to clarify ideas.

What does a discussion reviewing the structure and function of plants with a model built by students sound like?

Scan the QR code or visit *www.nsta.org/publications/press/extras/files/practices/questioning/video1.htm* to listen to a discussion with different types of questions.

How does the water get to the leaf?

The blue marble shows the water moving up through the roots into the stem.

Connecting Questions and Learning: Structure and Function of Plants

When exploring the concept of structure and function during a unit on plants, students make connections to their world by observing specific details of real seeds, roots, stems, and leaves at home, on the school campus, and in a school garden (McGough and Nyberg 2013b). Students conduct investigations such as examining and labeling the parts of a pumpkin in the fall, observing and comparing different kinds of seeds from the garden, observing a sunflower plant go to seed at the end of its life cycle, and planting seeds.

A variety of learning experiences involving plants give students context to engage in thinking and questioning throughout the unit of study (McGough and Nyberg 2013a). Reading informational text in addition to making firsthand observations stimulates even more questions. The teacher asks, "What questions do you have about plants and how they work?" This question causes students to reflect on what they have learned so far and then extend their thinking.

Using Unit Planning Guides

Student questions prompt further investigations, which advance the cycle of learning. As you design a unit, a planning guide can help you determine engaging questions, purposeful investigations, and authentic assessments to push the cycle forward. Students' extensive studies allow for crosscurricular connections. For example, after hands-on learning about plants, students might read informational text that describes and explains key ideas (English language arts standards) about how plants work (science content standards). Then, they could investigate how different variables affect plant growth (water, soil nutrients, and sunlight). Purposeful investigations help students build understanding of key concepts and might lead to an authentic performance task of building a model plant (science and engineering practices) to articulate how the structure of a plant helps a plant function (Figure 1.3).

Figure 1.3. Powerful Practices Model: Structure and Function of Plants
An example of the Powerful Practices model filled out during a unit on the structure and function of plants.

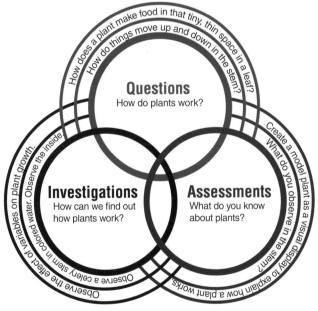

A comprehensive unit planning guide includes such crosscurricular possibilities (Figure 1.4, p. 14, illustrates a visual reference for crosscurricular connections) as well as content and academic vocabulary, resources, and differentiation strategies. An example of a complete unit planning guide for the unit on the structure and function of plants is shown in Figure 1.5 (pp. 15–17).

Figure 1.4. Brainstorming Crosscurricular Connections

Science: Investigate how seeds, roots, stems, and leaves work.

Technology: Produce and publish writing and collaborate with others through a classroom blog.

Engineering: Build a model plant to show how a plant works.

Math: Measure plants growing in the garden.

ELA: Record plant observations in a science journal including labeled drawings.

Social Science: Locate where foods are grown and transported to and from on the map.

Art: Observe leaf shapes and veins. Create a crayon resist of leaves.

Figure 1.5. Unit Planning Guide: Structure and Function of Plants

Timeline

Unit Planning Guide

Core Idea/Topic: Structure and Function of Plants

Concepts: Plant parts, plant needs, photosynthesis

Questions to Drive the Inquiry

1. What do you know about plants?

2. How do plants work?

Standards

NGSS

LS.A Structure and function of plants; LS.B Growth and development of organisms (plants); LS.C Organization for matter and energy flow in organisms (plants); LS.D Information processing; ETS1.2 Developing and using models

CCSS ELA

RI.1 Key ideas and details; RI.4, RI.5 Craft and structure; RI.7 Integration of knowledge and ideas; W.5, W.6 Production and distribution of writing; W.7, W.8 Research to build and present knowledge; SL.1, SL.2, SL.3 Comprehension and collaboration; SL.4, SL.5, SL.6 Presentation of knowledge and ideas

Student Questions

1. How does a plant make food in that tiny, thin space in the leaf?

2. How do things move up and down in the stem?

3. How does air go in and out?

Investigations

1. Observe how different variables affect plant growth (water, soil nutrients, sunlight).

2. Observe a celery stem in colored water.

3. Use straws to suck up water from a cup. Place a finger over the straw when it is in the water. Discuss.

4. Place a plastic bag over a leaf on a plant outside. Observe over time.

Performance Assessment

1. Students will build a model plant using straws, tubes, lids, netting, bubble wrap, and other objects.

2. Students will present their model plant to the class and explain how a plant works.

3. Students will write a report or create a brochure that explains how plants make food, what plants are used for, and why plants are important.

Figure 1.5 (continued)

Science	Technology	Engineering	Mathematics	English Language Arts	Social Science	Art
• How do the parts of a plant work? • How does climate affect the growth of plants? • How do seasons affect the growth of plants? • How does weather affect the growth of plants? • How does soil quality affect the growth of plants?	• Digital microscope: View seeds, roots, stems, and leaves. • Kidspiration: Organize key details in preparation for writing. • Class blog: Report data and information collected from plant experiments (include pictures). • Apps: Kids Discover: Plants; Leaf Snap.	• Build a model plant to show the parts of a plant and explain how the parts work.	• How would you measure a sunflower or other plant as it grows? • Create charts and graphs to organize the data collected from plant experiments.	• Read informational text to explain key details about plants. • Record observations and describe relevant details in journals. • Label diagrams and drawings. • Write reports communicating understanding of how plants work.	• Where do plants grow? • Research and compare plants that grow in different ecosystems. Locate geographic regions on a map. • How is food transported locally and globally?	• van Gogh: Observe, draw, and paint sunflowers. • Cézanne: Observe, draw and paint still life (using real vegetables and fruits). • O'Keeffe: Observe, draw, and paint flowers.

Learning Styles

Visual: See It
- Real plant experiments
- Campus field trips
- Time-lapse videos
- Picture books

Auditory: Hear It
- Discussions, large and small group
- Student collaboration
- Video presentations
- Audio recording

Kinesthetic: Do It
- Campus field trips
- Gardening tasks
- Building model plants

Differentiation Strategies

- Use songs and chants about plants.
- Use videos, technology applications, and hands-on investigations.
- Create charts that emphasize key vocabulary in context with photographs or drawings.

Figure 1.5 (*continued*)

Content Vocabulary

absorb, air, bloom, fall, flower, garden, grow, leaf, life cycle, nutrients, parts, petals, phloem, photo-synthesis, plant, plant root hairs, root, seasons, seed, seed coat needs, soil, spring, sprout, stem, summer, sunlight, veins, water, winter, xylem

Academic Vocabulary

change	display	information	present
clarify	evidence	journal	question
compare	explain	label	report
connect	facts	measure	support
contrast	ideas	model	report
describe	illustrate	observe	text
different			thinking

Resources

Butterworth, C. 2011. *How did that get in my lunchbox?* New York: Random House.

Cherry, L. 2003. *How groundhog's garden grew.* New York: Scholastic.

Fowler, A. 2001. *From seed to plant.* Danbury, CT: Children's Press.

Gibbons, G. 1993. *From seed to plant.* New York: Holiday House.

Kalman, B. 2005. *Photosynthesis changing sunlight into food.* New York: Crabtree Publishing.

Kudlinski, K. 2007. *What do roots do?* New York: Cooper Square Publishing.

Levenson, G. 1999. *Pumpkin circle.* Berkeley, CA: Tricycle Press.

Stevens, J. 1995. *Tops and bottoms.* Boston: Harcourt Brace.

Tagliaferro, L. 2007a. *The life cycle of a pine tree.* Mankato, MN: Capstone Press.

Tagliaferro, L. 2007b *The life cycle of a sunflower.* Mankato, MN: Capstone Press.

Project Learning Tree (*www.plt.org*)

Apps: Kids Discover: Plants (*www.kidsdiscover.com/shop/issues/plants-for-kids*) and Leafsnap (*http://leafsnap.com*)

Reflection

Very engaging unit for students! Students built one model plant as a whole class, working in small groups. Students would benefit from building their own smaller version of a model plant while collaborating with peers at tables. More students would actively participate in the thinking and learning. Find more experiments to illustrate how plants work. Some experiments need to occur over time. It may be helpful to begin experiments before beginning the unit.

Developing Questioning Strategies

What Types of Questions Do I Need to Ask, and When Should I Ask Them?

Questions serve many purposes. They help students connect concepts, think critically, and explore topics at a deeper level. They help teachers check for understanding and uncover student misconceptions. Questions can be used to clarify and to probe. Questions can extend students' thinking by requiring the students to justify their answers. Most important, questions involve students in the learning and cause the students to continue thinking and making connections even after the initial discussion ends (Table 1.2).

Teachers can ask several types of questions. Two main types are convergent and divergent. To check for understanding, the teacher asks a *convergent question* with one specific answer. To open up and expand the discussion to many possible responses, the teacher asks a *divergent question* with many possible answers. The questions define the focus of the learning. A discussion with only convergent questions feels like a game show, but a discussion with only divergent questions lacks direction. Discussions become dynamic when a blend of different types of questions is thoughtfully used. When deciding the types of questions to ask, ask yourself these questions:

⇨ What do you want to know?

⇨ How do you want your students to get involved in the learning?

Prior to the discussion, focus on the core idea (e.g., Structure and Function of Plants). Next, prepare questions that will engage student thinking. Think through the concept as if you were a student. What are possible misconceptions? What might the students wonder? How might the students need to justify and extend their answers?

What does a discussion sound like using these types of questions?

Scan the QR code or visit *http://www.nsta.org/publications/press/ extras/files/practices/questioning/video2.htm* to listen to a discussion with different types of questions.

For example, start with a divergent question: "What do you know about plants?" A student might respond, "Roots grow underground." Pose a *clarifying question*, asking for a yes or no answer: "Are the roots on a tree the same as the roots on a carrot?" Then, follow up with *probing questions* such as, "Why do you think that? How? Explain what you mean." The discussion encourages students to examine concepts at a deeper level. The teacher stimulates learning experiences with purposeful questions, and the students also learn to ask thoughtful questions with the purpose of wanting to know more.

Table 1.2. Types of Questions

Question Type	Question Purpose	Teacher Questions
Divergent (Multiple answers)	• Open-ended questions may determine prior knowledge, misconceptions, and possible areas to investigate.	• What do you know about plants? • What do you know about animal life cycles?
Convergent (One correct answer)	• Closed-ended questions check for understanding. • Review concepts.	• Where are the roots? • What are the stages of a chick's life cycle?
Clarifying	• Describe ideas in more detail. • Explain ideas in a different way.	• How do roots grow? • How does the chick hatch from the egg?
Probing	• Explain reasoning and deepen understanding. • Analyze ideas. • Compare and contrast.	• Are the roots on a tree the same as the roots on a carrot? • What if the chick egg is cracked before it is ready to hatch?
Justifying and Extending	• Hold the learner accountable for their thinking. • Providing evidence requires the learner to support and extend their ideas.	• Why do you think that? • What evidence supports your idea?

What Is Wait Time?

Wait Time 1

How long do you wait from the time you ask a question to the time you expect a student to answer?

Wait Time 2

How long do you wait from the time a student stops speaking to the time you follow with a comment or question?

According to Mary Budd Rowe (1986), the average wait time is less than one second. Wow! Imagine the rapid-fire expectations teachers have for students to respond. One second is not a lot of thinking time.

Rowe suggests that when students are given wait time of *three seconds or longer*

⇨ the length of the student responses increases,

⇨ students ask more questions,

⇨ students show more evidence of attending to each other,

⇨ failures to respond decrease,

⇨ a greater number of students participate, and

⇨ achievement improves on written measures where the items are cognitively complex.

Wait time, like any teaching technique, needs to be determined by the teacher. Some questions may need more time for deeper thinking; some learners may need more time to formulate their response. Wait time is an important factor to consider in leading a discussion. Always respect the learners. If a student doesn't know a response, no amount of wait time will help; however, if given a little more wait time—before and after the response—the discussion may be richer as students build on each other's ideas and gain a deeper understanding of the subject.

What Is Depth of Knowledge?

Webb's Depth of Knowledge (DOK) framework provides a common language to determine the cognitive demand or rigor intended by the standards and assessments. It consists of four levels, each increasing the depth of student understanding. The framework has been used to promote classroom discourse at higher levels of cognitive demand that includes knowledge of science content and processes (Webb 2005).

Teachers may use the DOK framework to plan questions and academic tasks that challenge students to think at higher levels, preparing them for real-life problem solving as well as complex assessments. Purposeful investigations offer many opportunities for student thinking and learning at the higher DOK levels.

For example, one teacher planned to have the students investigate the life cycle of chicks. She had collected informational text from the school library, and the incubator was set up and ready to go. The children anxiously awaited the arrival of the chick eggs being shipped from a hatchery. When the teacher opened the box, she discovered that many of the eggs had cracked. Rather than being discouraged, she realized that the unfortunate shipping damage presented a real-world learning opportunity. Questions about the shipping process led to authentic science and engineering investigations that explored all four DOK levels! (See Table 1.3 and Figures 1.6–1.8 [pp. 23–27].)

Student Questions to Investigate

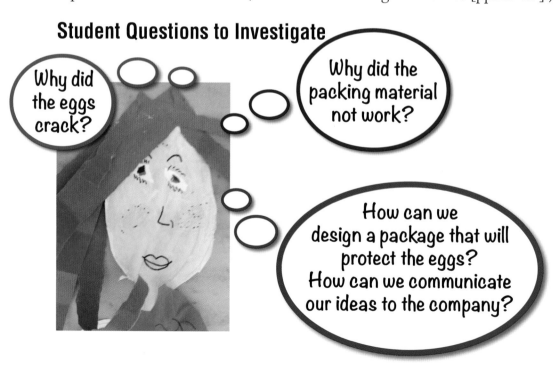

Why did the eggs crack?

Why did the packing material not work?

How can we design a package that will protect the eggs? How can we communicate our ideas to the company?

Figure 1.6. Depth of Knowledge: Growth and Development of Chicks
Using the DOK framework to study the growth and development of chicks.

DOK Level 1: Recall and reproduction

- Requires recall of information such as a fact, definition, term, or simple procedure, including following a simple process or procedure.

- Typically requires one step. Example: Recall the fact(s).

- Question: What are the stages of the chick's life cycle? Name the stages of the chick's life cycle in developmental order.

DOK Level 2: Skills and concepts

- Includes mental processing beyond recalling or reproducing a response.

- Requires more than one step. Example: Make observations. Infer explanation.

- Questions: What could have happened during shippping to damage the eggs? Describe the eggs and the container. Examine the carton that the chick eggs came in and the materials used for shipping.

DOK Level 3: Strategic thinking and reasoning

- Includes cognitive demands that are complex and abstract with more demanding reasoning.

- Requires multiple steps.

- Questions: How could the shipping package be modified to improve the condition of the eggs that are delivered? Design, build, and test the model and justify why it was successful or unsuccessful.

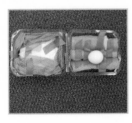

DOK Level 4: Extended thinking

- Requires high cognitive demand using higher-order thinking processes such as analysis, synthesis, and reflection; involves very complex ideas crossing multiple content areas.

- Requires multiple steps over time.

- Questions: How will you communicate your findings and persuade the company to modify its current packaging design? Write to the company persuading them to improve packaging design (may include a prototype).

Table 1.3. Standards-Based Learning: Growth and Development of Chicks
Examples of standards used during the study of the growth and development of chicks
DOK = Depth of Knowledge (see p. 22); ELA-English language arts.

National Standards	Standards-Based Learning
NGSS: Life Science **LS1.B:** Growth and Development of Organisms **NGSS: Engineering** **ETS1.1:** Asking Questions and Defining Problems **ETS1.2:** Constructing Explanations and Designing Solutions **CCSS ELA: Writing** **W.6:** Use technology to produce and publish writing as well as to interact and collaborate with others. **W.7:** Conduct short research projects that build knowledge about a topic. **W.8:** Recall information from experiences or gather information from print and digital sources. **W.9:** Draw evidence from informational texts to support analysis, reflection, and research **CCSS ELA: Speaking and Listening** **SL.1:** Engage effectively in a range of collaborative discussions, building on others' ideas and expressing their own ideas clearly.	**NGSS** Students observe that animals have unique and diverse life cycles (DOK Levels 1 and 2). Students ask questions about how the packaging and container could be modified to improve the condition of the eggs that are shipped from the hatchery (DOK Level 3). Students design containers to test possible solutions to the problem and communicate findings to the hatchery (DOK Level 4). **CCSS ELA** Students use tablets and computers to conduct research and build knowledge of packaging and shipping processes. Students interact, collaborate, and pose questions to further discuss design solutions as they test models. Students recall information from experiences and resources to write letters to the hatchery offering suggestions to improve the shipping process of live chick eggs.

Figure 1.7. Powerful Practices Model: Constructing Explanations and Designing Solutions

An example of using the Powerful Practices model to investigate the growth and development of chicks based on *NGSS* ETS1.2: Constructing Explanations and Designing Solutions

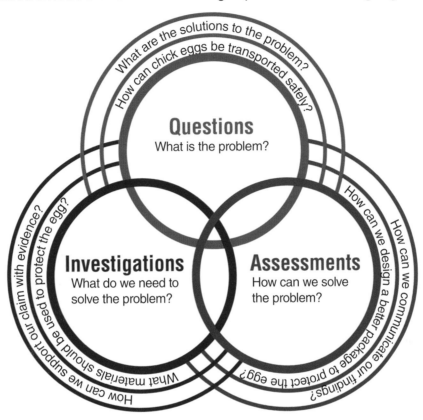

What are the solutions to the problem?
How can chick eggs be transported safely?

Questions
What is the problem?

How can we support our claim with evidence?
What materials should be used to protect the egg?

Investigations
What do we need to solve the problem?

Assessments
How can we solve the problem?

How can we design a better package to protect the egg?
How can we communicate our findings?

Teacher Metacognition

What is metacognition?

Metacognition: An awareness or analysis of one's own learning or thinking processes (Merriam-Webster 2011)

What is the difference between asking questions and defining problems?

The National Research Council (NRC 2012) makes a distinction between **asking questions** (scientists) and **defining problems** (engineers).

Scientific questions often arise from curiosity and wonder related to how the world works.

Engineering questions emerge to define a problem and develop a successful solution.

Early in the chick unit, the students were asking questions such as, "What are the stages of the chick life cycle? How does the chick hatch from the egg?"

Asking questions turned to defining problems when the eggs arrived broken. The broken eggs offered an unexpected teachable moment. By taking advantage of this opportunity, the teacher engaged her students in solving a real-world problem: "How can we design a better package to protect the eggs?"

The poor packaging and broken eggs led to student questions and higher levels of thinking (DOK Levels 3 and 4).

Figure 1.8. Engineering in Action!

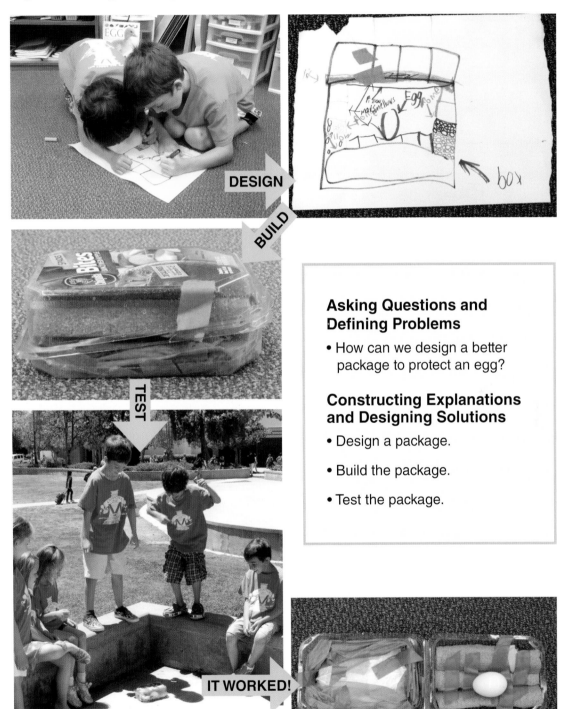

Asking Questions and Defining Problems

• How can we design a better package to protect an egg?

Constructing Explanations and Designing Solutions

• Design a package.

• Build the package.

• Test the package.

PART 2
How Do I Prepare for the Power of Questioning?

Part 2

Engaging Students and Teachers

How Do I Prepare for the Power of Questioning?

Involving students in the Powerful Practices model requires that you know your students as individuals—each student's way of understanding the world and way of participating in discussions—and as a group. This will help you create engaging questions and a collaborative environment that supports dynamic discussions, leading to purposeful learning that is applicable to the child's world. In a collaborative learning environment, the students understand that *all* students contribute to learning, *all* thinking and learning are valued, and *all* learners receive support. Each child knows his or her ideas will be valued just as much as peers' ideas!

Getting to know students and how they think helps teachers build on the qualities of the learners when developing a discussion. This is different from knowing which students struggle with reading or excel at math and who needs to sit near the teacher or is easily distracted. Understanding thinking characteristics helps you know how your students think and analyze information.

30

Who Are My Students, and How Do They Think?

Students come to the classroom as individuals with unique experiences. Listen to students as they collaborate with peers; wonder alongside them as they work independently. As you watch your students, you will gain insight into why students think what they think or how they connect different ideas. Insight helps teachers fuel or propel dynamic discussions by calling on a specific child at key moments when that individual child's thinking will contribute most to the discussion.

Holding a class discussion is very much like conducting an orchestra. The conductor needs to know when to bring different instruments in to create a specific sound or melody that conveys the message of the musical piece. Before the conductor can do this effectively, he or she must first know what each instrument sounds like on its own. A teacher who knows how each student thinks can access different perspectives, reveal more information, or propel a new investigation by questioning different students throughout a discussion.

To have an in-depth discussion, a teacher cannot always call on the student who may know the answer first. All students can contribute to learning by sharing their unique views of the world. Questioning is an instructional strategy that allows children to participate and contribute while providing time for students at all levels to process ideas. Understanding how students think helps the teacher conduct the discussion to make connections between concepts, clarify abstract thoughts, cultivate new ideas, or launch further investigations. The Observation Checklist (Figure 2.2, pp. 36–37) can help you get to know your students.

Thinking About Analysis

Some students see the big picture *and* the component parts of a concept quickly. Don't call on a person like this too early in the discussion. He or she might take the conversation to the big idea too soon—before you have cultivated new ideas or perspectives from other students. Calling on this person at just the right moment can help students deconstruct and reconstruct concepts or transition to the next logical step.

"Estevan, what do you think is happening inside the egg?"

Thinking About Big Ideas

Some students see the big picture and may begin planning projects without conceptualizing all of the important details. Calling on this enthusiastic type of learner at just the right moment may help launch the next investigation!

"Lacee, what do you think we could make to help us understand this idea better?"

Thinking About Questions

Some students are full of wonder and naturally inquisitive. These students demonstrate a strong drive to explore and want to observe items closely. Calling on this kind of student at just the right moment may spark new questions or create an opportunity to investigate ideas further.

"Michelle, how is the life cycle of the ladybug you found different from the life cycle of a chick?"

Thinking About Perspective

Some students look at things differently or from a unique perspective. Calling on one of these students at just the right moment may help you dive deeper into the discussion or articulate connections between related concepts.

"Jadee, how does a chick come out of an egg differently than a caterpillar comes out of an egg?"

Thinking About Details

Some students pay close attention to organization, sequence, and precise language. Call on this type of person to clarify abstract thoughts or provide missing pieces of information when needed.

"McKenzie, how can we organize the information to show the life cycle of a chick?"

Thinking About Meaning

Some students listen carefully and perceive relationships well. They show awareness of context clues and like to clarify meaning. These students can help define words in context or synthesize the contributions of others.

"Michael, the text said that the hen sits on her clutch after she lays the eggs. What do you think clutch *means?"*

Thinking ... Who Needs a Box?

Some students make conceptual leaps and offer a unique problem-solving perspective! Students like this may visualize concepts before all the pieces are in place. Calling on one of these people at just the right moment may help lighten the discussion and add humor or open the door to new possibilities.

"Christine, chicks do not have teeth to chew their food. How will the chick eat?"

How Do I Provide Opportunities for *All* Students to Participate?

During discussion, some students may not speak or may be hesitant to share. There are a variety of reasons for this reticence. Some may be trying to articulate questions and answers with new content and new academic language. Others may be less proficient in English and processing the English language as well as the "language of science." These students may need more than just wait time to formulate answers. Phrase your questions to them in context with real objects, models, or visuals. Figure 2.1 shows types of questions that can be used for students at different stages of communication.

Figure 2.1. Four-Tiered System of Questioning to Support English Learners

Silent Period
A student is not speaking in class.

⇨ **Teacher:** "Show me the root." (Response: Pointing to the plant part)

Beginning Level
A student speaks one or two words at a time.

⇨ **Teacher:** "Is this the *root* or the *stem*?"

Intermediate Level
A student is beginning to share short sentences.

⇨ **Teacher:** "Describe the root of the plant."

Advanced Level
A student is using academic language to explain concepts.

⇨ **Teacher:** "Explain how the plant gets water."

What does a student response sound like using these questions put in context with a model?

Scan the QR code or visit *www.nsta.org/publications/press/extras/ files/practices/questioning/video3.htm* to listen to a student response with different types of questions.

Figure 2.2. Observation Checklist: How Do My Students Think?

The teacher needs to know his or her students, their interests, their learning needs, and how they think.

An **observation checklist** can help you analyze how students think so you can plan questions to keep the discussion moving.

Record observations of students as they work independently, work in groups, participate in discussions, and interact with assigned tasks. When a student displays one of the characteristics listed, write his or her name in the box. Add tally marks next to the name if students display characteristics multiple times. Record observations over a week during different kinds of interactions and tasks.

Tip: Use different-color pens to record observations, signaling one color for each type of interaction or task (e.g., red for group work, blue for hands-on task, etc.).

Thinking About Analysis
- Sees the big picture
- Sees the component parts
- Sequences ideas logically
- Solves problems eagerly
- *Can deconstruct and reconstruct concepts or ideas*

Thinking About Big Ideas
- Dives in without conceptualizing
- Shows initiative and is a self-starter
- Is eager, enthusiastic, sometimes impulsive
- *Is always ready to launch the investigation*

Thinking About Questions
- Is naturally inquisitive
- Is full of wonder
- Has an imagination without limits
- Has a strong drive to explore and know
- *Is always willing to investigate further*

Thinking About Perspective

- Looks at things from a unique perspective
- Communicates ideas in a different way
- *Articulates connections between related concepts*

Thinking About Details

- Notices organization
- Pays attention to time and sequence
- Clarifies abstract thoughts
- Uses precise language
- *Contributes missing pieces of information when needed*

Thinking About Meaning

- Shows awareness of context clues
- Likes to clarify meaning
- Perceives relationships
- Observes and listens carefully
- *Synthesizes contributions of others*

Thinking Outside the Box

- Makes conceptual leaps
- Has a unique problem solving perspective
- Visualizes concepts before all the pieces are in place
- *Opens the door to new possibilities*

Building a Questioning Environment

How Do I Build a Collaborative Learning Community to Support Questioning?

A collaborative learning community includes both the *physical learning space* and the *cognitive environment*.

The physical learning space may be indoors or outdoors, real or virtual. The physical learning space in the classroom includes the investigation station. An investigation station is an organized learning space complete with resources and tools to invite students into a world of discovery. The investigation station includes resources such as a nonfiction library, vocabulary resources, and an observation area. The investigation station also includes instruments such as observation tools, data collection tools, and writing supplies.

The cognitive environment is carefully established as the teacher sets expectations for respectful interactions and dynamic discussions. Active listening plays a key role in modeling a learning community ready to engage in questioning. During discussions, the teacher models expectations as he or she looks at the speaker, leans forward, encourages the speaker with nonverbal signals (head nods, animated facial expressions), and asks purposeful questions.

Establish the cognitive learning environment to demonstrate value for children's ideas by asking, "What do you think? How do you know that? Tell me about your work or idea." Engage students in questioning and value each child's contributions as you model a sense of wonder and exploration.

You should also create an environment where mistakes play an expected and accepted part of the learning process. Actively model lifelong learning skills, including making

Teacher Metacognition

How do I help students learn the language of questioning to interact respectfully and to extend thinking?

Students learn how to interact in a discussion through modeling, teacher think-alouds, and sentence frames.

Modeling: A teacher may say something like, "Michael, that is an excellent question. I don't know the answer. We need to do some more research. I think I know of a book in the library that might help us."

Teacher think-aloud: "I think we may need to find some resources to help us understand the parts inside this bird of paradise stem and how it works. Lukas, please bring the tablet from the investigation station."

Sentence frames: Sample dialogue posted on charts may be helpful when students begin learning the skills of questioning. The charts can be taken down as the students build confidence and become more skilled in academic discourse.

"I agree with _____ because _____."

"I disagree with _____ because I was thinking _____."

 **Part 2**

mistakes. Finally, build on authentic and relevant connections as students learn to understand the importance of asking questions and engaging in dynamic discussions.

A collaborative learning community includes setting expectations for dynamic discussions (Table 2.1). Think about the arrangement of students during discussions and how students interact. For example, students may sit on the floor in a circle, in desks, or at tables. Sitting in a circle on the floor offers students the opportunity to see each person as they speak and encourages group participation. When desks are arranged in groups, consider strategies to focus student attention if student chairs do not face the speaker.

Table 2.1. Setting Expectations for Dynamic Discussions

Seating Room Arrangement	Respect for Self and Others	Nonverbal Communication Signals	Possible Types of Interaction
• Sitting in a circle • Small group discussions at tables • All students facing forward • Assigned seating vs. choice seating	• One person talks at a time • Everyone listens • Participants are respectful of each other's learning • Participants support each other • Dynamic community environment	• Raising hand to speak • *C* on forehead to signal making a connection • Holding two fingers up to signal quiet • Thumbs up to agree • Thumbs down to disagree	• Asking clarifying questions • "I agree because …" • "I disagree because …" • "I have a connection to what ___ said." • "I understand what ___ said but I also think …"

What does a collaborative learning community discussion sound like?

Scan the QR code or visit *www.nsta.org/publications/press/extras/files/practices/questioning/video4.htm* to listen to a discussion with different types of questions.

Expecting students to respect others helps students to listen and take turns speaking so that all students have an equal opportunity to participate. Signals may be used to help students communicate that they have connections or want to ask questions while another child speaks. For example, students may place their hands in the shape of the letter *C* to signal that they have connections to share (Figure 2.3). Using a nonverbal signal allows the teacher to acknowledge the child and give him or her an opportunity to speak next.

Unfortunately, there may not be time to hear every connection. The teacher may say, "Hold on to your thought. We will share more connections this afternoon." Or, "I love that so many of you have made connections! To make sure everyone gets to share their ideas, please write your connection on a sticky note and place it on the chart. We will look forward to reading each other's ideas."

Figure 2.3. C for Connections
Students place their hands in the shape of a *C* on their foreheads to signal they have connections.

How Do I Organize Resources to Engage *All* Learners?

Organizing resources for student accessibility engages students in a variety of opportunities for authentic literacy development. Students seek information, dig deeper, and share ideas with others when the environment is prepared and organized. The investigation station (Figure 2.4) provides a hub for focused study. Students apply standards-based learning through active engagement, reading text, writing, making connections, analyzing text and information, and engaging in questioning.

Figure 2.4. Investigation Station

When you provide students with technology, you enhance and support their ability to make observations (using a hand lens or microscope), document evidence (recording videos on a phone or tablet), and communicate their findings (digitally displaying collected data, multimedia presentations, etc.). Technology resources may extend the learning beyond the walls of the classroom (e.g., classroom blog or video conference with pen pals). Teachers may use apps and internet tools to provide support resources for investigations and assessments (Figures 2.5 and 2.6, pp. 44–45). Each teacher may design the investigation station to address a variety of student needs and learning styles by including materials for observations, investigations, and assessments. Students construct questions as they use the investigation station materials: nonfiction library, vocabulary resources, observation area, observation tools, data collection tools, and writing supplies.

For example, when students observe different types of seeds, the investigation station may include the following:

1. **Nonfiction resources:** For example, *The Life Cycle of a Sunflower* (Tagliaferro 2007) and *From Seed to Plant* (Fowler 2001).

2. **Content vocabulary:** For example, labeled pictures of flowers and their seeds.

3. **Observation items:** For example, sunflower head, sunflower seeds, hollyhock seeds, marigold seeds, seedpods.

4. **Observation tools:** For example, hand lens, microscope, and small containers to hold seeds.

5. **Writing tools:** For example, science journals, colored pencils, pencils, Venn diagram templates to compare and contrast seeds.

Marigold seeds

Figure 2.5. Organize Resources

How to organize resources at an investigation station.

Nonfiction Library

- Provide a focused area for a different type of reading. Materials may be gathered from the school or public library, teacher collections, internet resources, or materials brought in by students.

- Designate an area for informational text including books, magazines, and other text resources to invite students to read, research, and investigate.

- Organize informational text into categories for easy student access. Possible categories: Animals, life cycles, habitats, Earth science, physical science, science magazines.

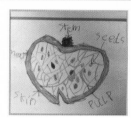

Vocabulary Resources

- Integrate vocabulary into student work displays.

- Label items on display for observation (e.g., experiments, investigations, animal study).

- Organize vocabulary in pocket charts or displays to illustrate key words in context (e.g., seed, roots, hair roots, stem, leaves, veins, flower).

- Present terminology in context (with pictures or actual objects).

- Introduce terminology during learning experiences to maintain connection to the content.

- Create an active word wall for new academic vocabulary and content vocabulary as it is introduced for student reference.

Observation Area

- Organize and label items to observe.

- Display relevant books with real objects.

- Items may be placed by a teacher or brought in by students from recess or home. If you are studying plants, bring in plants. When you cannot bring in the items, use pictures and technology resources.

- Example: Sunflower head, sunflower seeds, sprouting sunflower plant, and books on sunflowers; pinecone, bark, sap, pine needles, books on pine trees.

Figure 2.6. Organize Tools

How to organize tools at an investigation station.

Observation Tools
- Organize and label tools for observation.
- Observation tools should be organized and available for students to use each day. Specifying a place in the classroom to house these items offers a learning space for purposeful work.
- Possible items: Hand lens, petri dish, tweezers, microscope, digital microscope, and small containers with air holes (for animals that visit the classroom).

Data Collection Tools
- Organize and label tools to collect data for easy access.
- Students may find items to measure or compare on the playground or bring in interesting finds from home. These items serve a learning purpose when tools are available to use each day.
- Possible tools: Rulers, scales, measuring tape, computers, cameras, and tablets.

Writing Supplies
- Organize and label supplies to record observations.
- Students may have individual supplies, but additional supplies should be organized and available for student projects and investigations as they arise.
- Possible supplies: Paper, folders, paper trays, pencils, colored pencils, glue sticks, scissors, journals, computers, and tablets.